A Pilot's View of Flying Machines

1900-1909

By Ed Middleton

A Pilot's View of Flying Machines

1900 - 1909

By Ed Middleton

ISBN: 978-0-9669400-6-0

Edited By Jean F. Middleton

This book is dedicated to my wife, Jean. She has been by my side since 1973, putting up with my nonsense for nearly forty years.

What more could one do to deserve a dedication?

I love you, Jeanie.

Ed

CONTENTS

Chapter 4. Demonstration Flights and Contracts

Chapter 5: Flying Machines 1908 - 1909

Epilogue: Events Following 1909

CHAPTER ONE

First Flying Machines

The story of the first flying machines is filled with fascinating characters. The following are some of the most important, all of whom never doubted that man could and would fly:

Orville Wright - 1871-1948 - First Flyer

Wilbur Wright - 1867-1912 - Second Flyer

The following three men supplied information to the Wright brothers on various aspects of flight. Dr. Samuel Langley, who tried valiantly to become the first to fly a powered man-carrying machine himself, failed when his rear wings fell off as his plane was launched. So Orville Wright has the honor of being the first pilot.

Octave Chanute
1832- 1910
Railroad Designer
His book on aerodynamics, *Progress in Flying Machines* was published in 1894.

Samuel Langley
1834 – 1906

Third Secretary of Smithsonian Institute
His steam-powered flying machine crashed in December
1903.

Otto Lilienthal
1848-1896
A German Engineer

Wright brothers studied his book
*Practical Experiments for the Development of Human
Flight, published in 1*893.

Otto Lilienthal built 25 different gliders, both
monoplanes and biplanes. He used body shifting to help
control his mild turns. He died in 1896 after a crash in
which his back was broken.

Lilienthal Gliding Off a Hill near Berlin

Early Experimentation by Wright Brothers

Wright Family

Wilbur and Orville Wright, who never married, had two married brothers, Reuchlin and Lorin, and a younger sister. The sister, Katherine (Kate), lived with her unmarried brothers and kept house for them. She did not marry until late in life. Their mother, Susan Koerner Wright, who was born in 1831, died at 59 of pneumonia. Kate, who was 15, then took her place in the Wright household. Their father, Milton Wright, a minister and theology professor (1828-1917) lived to see his sons' success.

First Gliders

Orville and Wilbur owned a bicycle shop in Dayton Ohio and had been working privately on gliders that would carry a man. Sometime in the early half of 1900, they wrote the Smithsonian Institute to request material on aerodynamics, and they received a great deal of information from the museum.

After the brothers received this information from the secretary of the Smithsonian, they were able to improve their gliders and made several large enough to carry a man. As it happened, the secretary, Samuel P. Langley, was himself very much interested in flying gliders.

The Wright brothers had started flying gliders even before 1900. They required a location, however, that had better wind conditions than Dayton, Ohio. They wrote to the U.S. Weather Bureau and received information on wind conditions throughout the United States. They chose a sandy beach about 60 miles north of Cape Hatteras, North Carolina. That location, called Kitty Hawk, is just north of Kill Devil Hill, an old Coast Guard station.

The brothers first took their current glider apart for shipping to Kitty Hawk and shipped the entire glider, with the exception of the right and left main spars that were too long for normal shipping. They knew they could purchase new spar lumber at Norfolk, Virginia. The glider had wings that were 40 feet long, and it would not be reasonable to transport a 20-foot piece of wood that far.

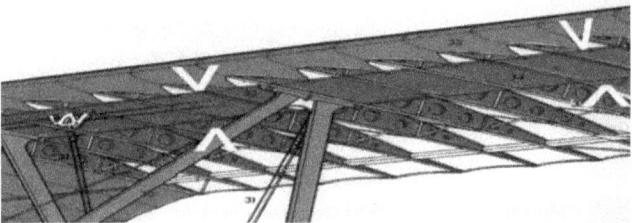

Top Wing of the Fokker Triplane of WWI

The main spar is the long single wood beam (between the yellow pointers) that runs from the wing tip to the central section of the plane. It essentially carries the entire wing load to connect with the fuselage. In this particular WW I aircraft, the gray struts hold this third top wing to the main body, or fuselage, of the aircraft.

Wilbur shipped all the glider parts to what is now Newport News, Virginia. On Thursday, September 6, at 6:30 p.m., he boarded a train in Dayton, Ohio. After changing trains three times on three different railroads, he arrived in Newport News. Wilbur then collected his glider parts and boarded a ferry that crossed Hampton Roads, arriving in Norfolk. There he was able to purchase two boards to be used as wing spars

He then traveled to Elizabeth City, Virginia, on Saturday to find a boat to carry him and his precious glider parts down the sound to the outer banks near Kitty Hawk. He could find only one boat captain who would take the job. The old beat-up vessel leaked badly, but fortunately the captain was great at bailing and keeping the boat afloat.

After much delay in repairing the boat, the group set sail on Wednesday in the vermin-infested schooner, and they finally arrived at Kitty Hawk Bay at 9 p.m. Wilbur arrived at Kitty Hawk for the first time on September 13, 1900. He spent the night onboard the schooner but was afraid to eat the food and ate only a jar of jelly that his sister had stowed in his bag.

Wilbur's trip was humorously described in a letter written by him to his family in Dayton. The family all agreed that this was the funniest story Wilbur ever wrote.

Wilbur Flying a Glider at Kitty Hawk in 1900

Original Tent on Left – The Wright brothers built their
new work shed in 1901

The Wright brothers returned to Kitty Hawk each September from 1900 on, until they made their first powered flight with a man aboard in 1903.

Factors Leading to Success

During their first three years of serious glider flying and experimentation with the different aspects of flight, the brothers kept careful records. Understanding five essential factors led to their successful powered, man-controlled flight as described in the following paragraphs. These five factors are:

The Shape of the Wing (airfoil) (1)

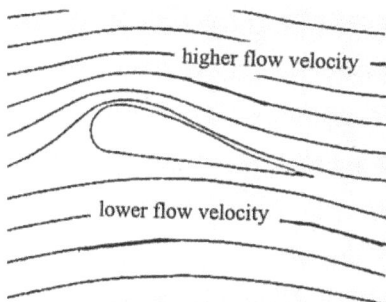

This is based on the Bernoulli principle.[1] If you think of the air as a fluid, the air going over the wing has to speed up to arrive at the same spot as the air going under the wing. The result is that the pressure on top of the wing is lower than that under the wing, allowing the wing to rise or lift. Today we have the *Boundary Layer effect,* but it is really just an expansion on the law of conservation of energy as shown by Daniel Bernoulli.

---

[1] **The Bernoulli Equation** can be considered to be a statement of the conservation of energy principle appropriate for flowing fluids. The qualitative behavior that is usually labeled with the term "Bernoulli effect" is the lowering of fluid pressure in regions where the flow velocity is increased. This lowering of pressure in a constriction of a flow path may seem counterintuitive, but seems less so when you consider pressure to be energy density. In the high velocity flow through the constriction, kinetic energy must increase at the expense of pressure energy.

Professor Lilienthal's Gliders
It was from his book that the Wright brothers first learned
about lift. Notice how the wings are arched, forming an
airfoil giving lift to the glider

Rudder (2)

The second important factor learned by the Wright brothers was the structure at the tail of the aircraft, called the rudder, that was used to effect horizontal changes in course.

Elevator (3)

The elevator is a movable control surface usually attached to the horizontal stabilizer of an aircraft that is used to produce motion up or down.

Wing Warping (4)

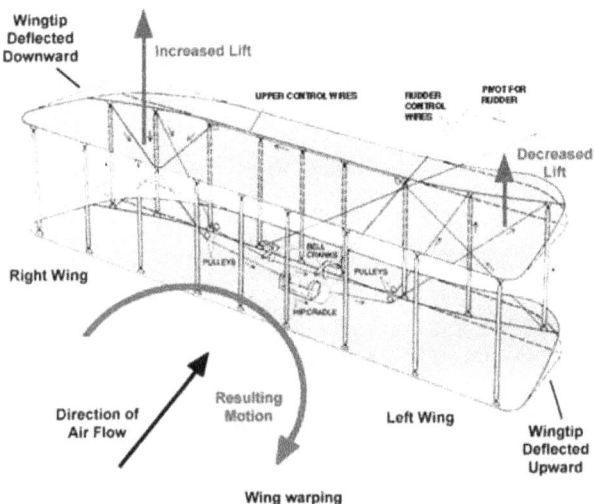

Wing warping

The simple explanation of wing warping is that the two wingtips are twisted in opposite directions to increase the lift on one wing and decrease that of the other. The example shown above illustrates one of the Wrights' planes as viewed from the front.

In this example, the trailing edges of the two right wings are twisted downward from the original position indicated by the dashed lines. Meanwhile the trailing edges of the two left wingtips are twisted upward.

The effect of this twisting action is that the angle of attack seen by the right wingtips is increased when compared to the initial position of the wings, and the two left wingtips experience a lower angle of attack.

As the angle of attack increases, lift also increases, so the right wings create more lift than the left wings. This difference in lift causes the aircraft to roll, right wings up, left wings down.

In the drawing on the previous page, the elevators and rudders are not shown. The Wrights' first flyer, side view shown on page 16, shows the twin elevators in the front of the machine and the twin rudders in the rear. When a turn is made, a little up elevator keeps the nose up and a little rudder helps point the craft in the correct direction.

Body Shifting (5)

The Wright brothers learned that by shifting their weight, they had at least slight control when, for instance, they needed to climb a bit or turn just a little.

Now that they knew how to fly gliders, they had only to incorporate an engine powerful enough to carry their machine into the air and keep it there. The Wrights' gliders, as well as their first powered flying machine, were mounted on runners like skis used in the snow. They built a mildly sloping ramp to keep their machine straight until it became airborne.

Wright Brothers' Wind Tunnel

The Wright brothers designed their own wind tunnel and tested many, many shapes and parts to find the best of all shapes for each item used in their gliders. The final items needed to fly their first flying machine were an engine and a propeller. The images on this page show a

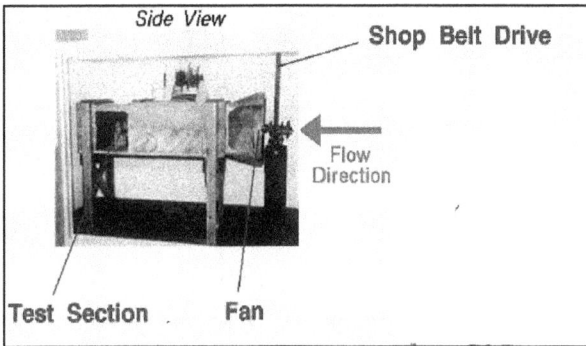

couple of views of their wind tunnel.

If you make a small wing and turn it facing the direction you wish to go, then it will pull you in that direction, and the propeller rises in the direction that the thrust is developed, forward in this case.

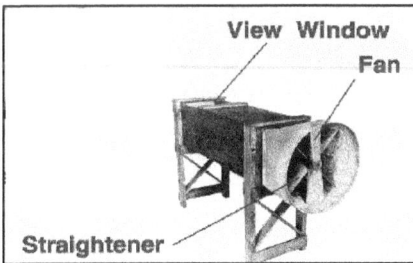

The Wright brothers certainly understood torque because they used two propellers, one turning in a clockwise direction and another turning in a counterclockwise direction.

Torque is thus eliminated by having one prop creating torque to the right and the other prop creating torque to the left, together giving a balancing effect.

Front View of the First Wright Flyer

Side View of First Wright Flyer

Note that the pair of elevators is in the front and the pair of rudders is in the rear.

The propellers behind the engine are normally called pushers, and the pilot lies down next to the engine facing the front of the craft, which is to the right of the photo.

Enter the Ghost

Augustus Herring was born in Georgia and attended the Stevens Institute of Technology in Hoboken, New Jersey, to study mechanical engineering but never graduated. His thesis on flight was considered "fanciful" by his mentors because his subject was supposed to be on marine steam engines, not flight.

Herring was an unpleasant man with a big ego who fancied himself as a great inventor in the field of aviation. His boasts were often designed to deceive others.

Augustus Moore Herring
1865 -1927

Augustus Herring appears many times in this story, sort of like a ghost.

His only real claim to fame is that he built a couple of gliders and met Octave Chanute in the late 1890's to discuss gliders. Chanute was invited to Kitty Hawk in 1902 by the Wright brothers because he had given them a great deal of advice. He had suggested, for example, that they build the wind tunnel for testing the shape of their glider wings.

Chanute asked permission from the bothers to bring another person with him. Of course they agreed, and so in 1902, Chanute and Augustus Herring arrived at Kitty Hawk with two gliders. Herring's gliders did not perform well, while the Wright brothers impressed their guests with the performance of their gliders.

First Powered Flying Machine Carrying a Man

On the morning of December 17, 1903, the first controllable machine climbed into the sky carrying a man. The time recorded by the photographer was 10:35 a.m.

This machine was designed by Wilbur and Orville Wright of Dayton, Ohio. They had chosen a good site where they might find the proper wind and a safe landing area. It was a sandy beach about 60 miles north of Cape Hatteras, North Carolina.

Their location, called Kitty Hawk, is just north of an old Coast Guard station Kill Devil Hill. Kill Devil Hill Coast Guard Station is mentioned because after the Wright brothers had set up everything for their flight, Orville went down to the station to gather some witnesses.

Five men from the Coast Guard station did accompany Orville to the launch site to witness the fun, good or bad. Orville asked John T. Daniels, one of the witnesses, to photograph the flight, but only if it did get airborne. All the men's names are listed in Orville's letter describing the event.

That day the brothers made four flights. Orville flew the first, which lasted 12 seconds and went about 120 feet. Wilbur flew the next flight, and after Orville had flown the third flight, Wilbur made the longest flight, remaining airborne for about 1 minute and flying a distance of 850 feet. (See photo below taken by John T. Daniels of the Coast Guard Station.)

Orville's First Flight

Up Pops the Ghost

The next time Herring appears in the lives of the Wright brothers was in 1904, when he found out that the brothers had applied for a patent for their powered flying machine.

Herring claimed that he held a prior patent on a machine similar to theirs. He offered to form a joint company to market the Wright flyer on the basis of one-third interest for him and two-thirds interest for them. The Wrights ignored what they termed as Herring's "rascality."

Wilbur wrote to Chanute about the Herring offer, including a copy of Herring letter. Chanute was not surprised because he had already determined that Herring was just a "blow hard."

The Competitor: Samuel P. Langley

Just 9 days before the Wright brothers made their first flight in an engine-driven flying machine, their nearest competitor, Dr. Samuel P. Langley, almost succeeded in flying a steam-engine-powered machine. The rear portion of the craft fell off, however, and the machine fell into the Potomac River, near Quantico, Virginia.

Samuel P. Langley and the Smithsonian Institute dispute

Langley, a graduate civil engineer, was asked to serve as an assistant to the second secretary of the Smithsonian Institute, Spencer F. Baird. Baird was in poor health and felt the stress of his job. Upon Baird's death in 1887, Langley was appointed to succeed him, becoming the third secretary of the institute, which was founded in 1857. Langley had been interested in flying machines for several years and had built six steam-powered aerodromes, as he called them.

They were unmanned gliders with propellers and used steam engines for power. He launched his aerodromes at Quantico, Virginia, from a scaffold that he had built on a barge anchored at the west edge of the Potomac River. All of his vehicles crashed into the river, with the exception of aerodrome number 5, an unmanned glider that actually made it across the river. Langley had influential friends in Washington who arranged for him to meet with President-elect Teddy Roosevelt.

Langley convinced many in Washington that he could build an aerodrome that would carry a man into the air. A bill was passed giving Langley $20,000 to complete his project. Although this was more than he requested, he would have to appeal many times for more money.

The Ghost Reappears

Not long after Augustus Herring had appeared at Kitty Hawk in 1902 with Octave Chanute, he was introduced to Langley at the Smithsonian. He told Langley that he knew a great deal about gliders and would be glad to work with him to develop a powered glider. He also told him that he knew all about the Wright brothers' gliders and that he was a friend of Chanute.

Langley, who also knew Chanute and had had many discussions with him about gliders and flying in general, wrote to ask him about Herring. By that time Chanute had figured out that Herring was full of hot air and was mostly talking through his hat about his knowledge.

Langley then told Herring that he just couldn't fit him into his project. Langley had spent about 3 years working on his man-carrying airdrome, as he called it. In October 1903 he tried out his newest invention. He did not try to fly it himself but had an engineer, Charles. Manley, pilot the aerodrome. The aerodrome had just been launched off the scaffold-like launch pad when the rear wing broke off, and the aerodrome plunged into the river. Fortunately, Manley was not critically injured, but Langley was so disappointed that he never tried to build another aerodrome. His government support was withdrawn, but this was not the end of the story. Even after

Langley's death in 1906, the Smithsonian Institute insisted that Langley had really built the first flyable man-carrying aircraft.

Pilot Manley upon Launch . . .

. . . before the wings fell off.

Wright Brothers' Second Attempt

About 10 days after their first flight on December 17, 1903, the brothers made a few adjustments to the controls that warped the wings and tried to make one more flight.

A sudden gust of wind, however, caused damage not only to a wing but also to their gasoline-powered engine. They then packed up and returned to their hangar in Dayton. Although they no longer had the exact flying machine that they had flown for the first time, they knew they had achieved the success they wanted.

Engine for the First Wright Flyer

The internal combustion engine was driving two propellers via belts that rotated in opposite directions. It was gasoline fueled and water cooled, with a magneto-generated electrical spark. It had 4 cylinders rated at 12 horsepower. Below are images of Wright Flyer No. 1 aircraft engines, Top and Bottom Quarters, respectively.

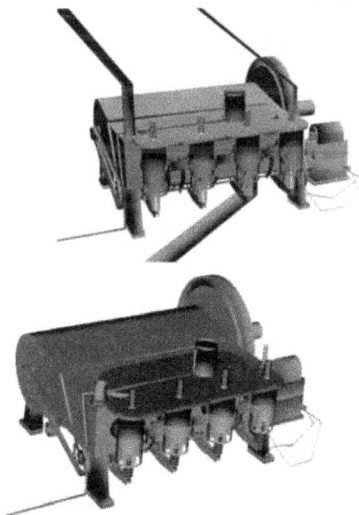

*Replica of Original Wright Flyer's
Engine Mounted on Plane*

The brothers actually built the entire aircraft with one exception. They hired a machinist-mechanic, Charles Taylor, who actually built most of the engine with a little help from the brothers. No such engine was affordable or available at that time. A breakdown of this engine is available in appendix A.

Wright Brothers' Costs Compared to Those of Langley

According to *To Conquer the Air*, a book by James Tobin, Langley's enterprise now approached $70,000. The Wrights figured up the total cost of their experiments from 1900 through 1903 including train and boat fare to and from the Outer Banks, at just under $1,000.

Secrecy of the Wright Brothers

Tobin explains the Wright brothers' apparent secrecy as follows:

Their aim was to sell everything, including their patents for $250,000. They wanted a contract that both parties would sign before the Wrights made a single demonstration flight. This condition would protect them from window-shoppers who wanted to see the machine with no real intention of buying. The buyers would be protected by a clause saying no money would actually be paid until the Wrights made fully satisfactory demonstrations. This condition was a hard one for governments to accept, as the brothers soon learned. But they would not back down from it easily. They were determined that no detail of their design would be stolen. The only way to secure their measure was to keep the design secret until their compensation was guaranteed. They had applied for patents but the patents had not yet been granted. And in any case, they believed that the secret of flight, if sold to the right party, would gain more of what they wanted than patents and manufacturing ever could—not just money but freedom, too. Better to get their price for the whole ball of wax—machines, designs, patents, and training—than to burden themselves with a patent war.

CHAPTER TWO

The World Takes Notice

Glenn Curtiss

Glenn Curtiss courting his fiancée in 1897

Glenn Curtiss, who was born in New York in 1878, bought his first bicycle at the age of 17 and became a champion bicycle rider at a major upstate New York bike-riding contest at 18. His father had died in 1882, so he must have been a self-made man because he married in 1898, opened his own bicycle shop in 1900 and in 1902 formed the G. H. Curtiss Manufacturing Company and started manufacturing bicycles and Hercules motorcycles.

Glenn Curtiss in 1907

Curtiss Motorcycles

1903 – Curtiss travels one mile in 56.4 Seconds

In 1903, the year of the first Wright brothers' flight, Curtiss set a world motorcycle speed record of 1 mile in 56.4 seconds, or a little more than 60 miles per hour. He raced in the Ormond Beach, Florida, time trials and set a new speed record in 1904.

In 1906 his company developed an 18-horsepower V4 engine, and Curtiss met the Wright brothers later that same year when he attended the Dayton, Ohio, Fair. They undoubtedly discussed flying machines because Curtiss started building his first flying machine in that same year.

In 1904, Curtiss set a ten-mile speed record.

In 1907, using a brand new 40-horsepower V8 engine, Curtiss set a new speed record on his motorcycle of 136 miles an hour at Ormond Beach, Florida.

In 1907, Curtiss became the fastest man in the world,
taking this motorcycle to 136.36 miles per hour

All the men mentioned in this chapter had an important part in the beginning of the flying machine, and important details are covered later as they come into play after a few years.

First Flying Machine Patent

In January 1904, just after their first flight, the Wright brothers applied for a patent in the United States, as well as in Great Britain, France, and Germany. Patents were granted in Great Britain and France in 1904. Germany granted a patent in 1906. The United States told the Wright brothers that their lawyer was not qualified to apply for a patent and that they needed to have a government-certified patent attorney make the application. A U. S. patent was granted in May 1906. Later in this account I cover some of the details of the lawsuit that the Wright brothers brought against Glenn Curtiss in 1908 about patent rights.

The first two pilots who flew a powered aircraft were Orville Wright, the very first, and a few hours later Wilbur Wright became the second pilot to fly the same machine.

Although the Wright brothers were the first to actually develop a man-carrying powered flying machine, the French were not far behind, and in fact there were many glider clubs in France, and Germany was close behind the French.

On October 14, 1905, L'Aero-Club de France (ACF) was formed. Aero Club America was also started in the United States about the same time. Glenn Curtiss won the first monetary prize sponsored by the American club and the *Scientific American Magazine* in 1908.

France and Aviation in the 1880s to 1903

Alberto Santos-Dumont

A Frenchman, Alberto Santos-Dumont, was well on his way to developing a flying machine to carry a man. Alberto Santos-Dumont was born in Brazil in 1873 into a wealthy family. His father sent him to France at the age of 18. During his years of education in France he became enamored with flying machines and first built a glider modeled after box-kite type gliders, developed by Lawrence Hargrove in France in the 1880s.

Lawrence Hargrove Glider

He soon got involved in building balloons that could carry a man and be steered in the direction desired. Alberto built several dirigibles (balloons that can be directed) in order to win a prize offered by L'Aero-Club de France (ACF) to be awarded to the first dirigible that could take off from Saint-Cloud, circle the Eiffel Tower, and return to Saint-Cloud.

On November 13, 1890, Alberto took off in his Santos Dumont No. 3 and circled a few times around the Eiffel Tower and then flew back to where he started. For this feat he was awarded the Deutsch Prize, which included 100,000 francs that he distributed to his workers and beggars of Paris.

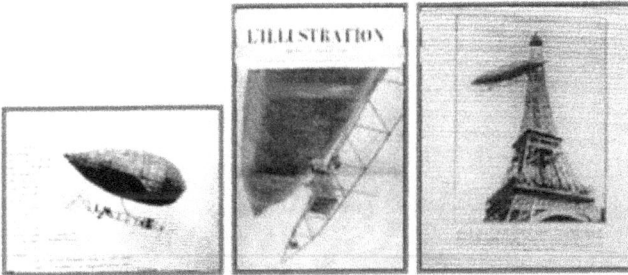

Three Views of the Santos Dirigible No. 3

Alberto Santos-Dumont in Basket of his No. 3

Second Manned Flight

According to Ernest Jones in his well-researched article entitled "The Very Earliest Early Birds," Alberto Santos-Dumont was previously thought to have been the first pilot after the Wright brothers to fly a powered flying machine; however, there were two others who flew a few months earlier in 1906. The third man to fly was also a Frenchman, Trajan Vuia, and so the second craft to fly was a miniature tractor monoplane with a 25-horsepower carbonic-acid gas engine. The flight took place in Sartrouville, France, on March 18, 1906.

Trajan Vuia flew his flying machine about eight times for short distances but crashed in July 1907 and never attained further prominence in aeronautics.

Third Plane to Fly - The Fourth Pilot

Above is a photo of J.C.H. Ellehammer in Lindholm, Denmark, who made his first flight of 42 meters on September 12, 1906. It was more like a helicopter and is credited with being the first plane in that class. This machine had a 9- horsepower engine. He did build other flying machines through 1908 but was not prominent after February 13 of that year.

Fourth Flying Machine

Alberto Santos-Dermont is now considered to be the man who flew the fourth flying machine one day after the Ellehammer flight on September 13, 1906, but the two flights listed above are in dispute according to many of his fans. It is undisputed, however, that he was the first to fly in France. Finally in 1906 he had his flying machine ready to fly. Here is a chronicle of the sequential events that took place when his test flights began:

Flight 1 July 1906 - The craft did not get off the ground (adjustments were made).

Flight 2 September 7 - The craft rose off the ground momentarily (more modifications).

Flight 3 September 13 - The craft rose to the height of 1 meter.

Flight 4 October 12 – The craft flew a distance of 6 meters.

Flight 5 November 12 – The craft flew a distance of 220 meters at the height of 6 meters and at the speed of 37.358 kilometers per hour.

Alberto Santos-Dumont was awarded the "Archdeacon Prize."

Alberto Santos-Dumont Fying His HTA (Heavier Than Air) Machine, he called 14-BIS. It was made of canvas and bamboo.

French enthusiasts, not realizing what the Wright brothers had already accomplished, hailed this event as the world's first successful flight of a powered heavier-than-air machine.

Santos-Dumont's Best Known Plane, La Demoisette, *Flew in 1910*

Wright Brothers' Flying Machines 1903-1905

On December 18, one day after their first flight in 1903, only three newspapers in the United States mentioned the Wright Brothers' flight, and their accounts were largely imaginative.

1903 Flyer 1 – Below is the Wright brothers' first powered aircraft; the first in which anyone made a sustained, controlled flight. As in their earlier gliders, it featured a variable-camber twin canard in front to control pitch and a twin rudder in back to control yaw. Roll was controlled by warping the wings.

1904 Flyer 2 – The following photo is the Wright brothers' second powered aircraft in flight, almost a copy of *Flyer 1*, with which they learned they still had a lot of work to do before they had a practical airplane. It was also the first aircraft on which the Wrights used their distinctive "bent-end" propellers.

Flyer 3, the next photo, is considered to be the world's first practical flying machine with many improvements over their first flyer. Their machines were still launched

and landed on sort of skis. That is why they did not attempt to win the first prize for taking off on wheels instead of runners. The 1905 aircraft was their third powered machine. Both the canard and the rudder were extended out from the aircraft to make it easier to control and to prevent the nose from dropping in a turn.

January 1905 - The Wright brothers contacted their congressman to inquire if the U.S. Government was interested in their experiments and machine. The reply was a form letter from the president of the Board of Ordnance and Fortifications indicating that the board was not interested in "financing experiments."

February 1905 – The British War Office sent a letter to the Wright brothers asking them to submit terms for the purchase of their "aeroplane."

October 1905 - At the urging of Octave Chanute, the Wright brothers again wrote to the U.S. Government offering their services to the Secretary of War. The reply was another rebuff.

Other Events of 1906

February 27 – Professor Samuel Pierpont Langley died at age 72. Although he is gone he is not forgotten as events show as the story progresses.

May 18 – Wilbur Wright sailed for Europe to complete negotiations for purchase of flying machines by Great Britain, France, Italy, and Germany.

May 22 – A U.S. Patent was issued to the Wright brothers.

This is a photo of U.S. patent number 821,393. The patent is all about aircraft control; it has nothing to do with power.

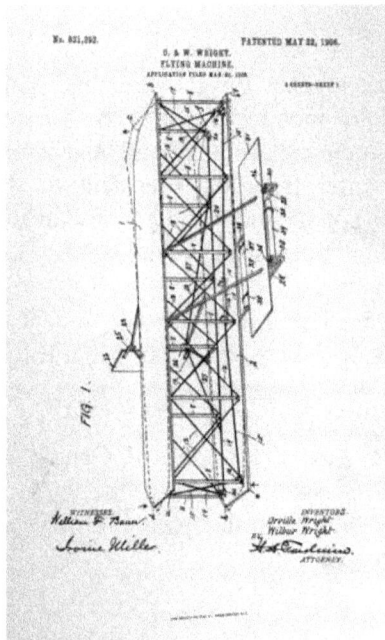

1907 - The Year the World Catches On

In France, the Voisin brothers had been building and selling gliders from their factory for many years. Sometime in 1906 a customer, Leon Delagrange, asked them if they could build a powered machine that a man could fly. They accepted the challenge.

Charles Voisin, left: 1882-1915
Gabriel Voisin, right: 1880-1973

Charles Voisin Demonstrating His Machine

Charles was the *sixth* man to fly a new model craft, and it was the *fifth* flying machine model that carried a man. This first flight was made on March 16, 1907. The brothers made improvements, and Charles flew it again in March before making a few more modifications and flew its last demonstration flight on April 8, 1907, before delivery to the customer. The customer did not fly the plane for quite some time after he bought it.

The Voisin brothers became a major manufacturer of flying machines for several years.

A Possible Romance: Another Voisin Story

On March 16, 1907, Charles Voisin became the sixth pilot to go up in a flying machine. Later on that same day he is credited with taking up a female passenger, Elise Raymonde Derroche. Sometimes she is called a countess and other times a baroness.

Elise Raymonde Derroche

There may have been some sort of romance here since they were together again in a high-speed auto crash on September 26, 1915, in which Charles was killed and the baroness injured.

Pilot Number 7, another Frenchman

Louis Bleriot
1872 – 1936

Below is a photo of his first flight is shown on the next page in a flight made at Bagatelle, France. He traveled five or six meters. The plane he used to cross the English Channel is shown near the end of the book.

Louis Flew This Weird-Looking Duck First on April 5,
1907

Pilot 8, Flying Machine 7

The next pilot to fly a different powered machine was Henri Farman (1874-1957), again a Frenchman.

Farman first flew this plane on September 30, 1907. The plane flew a distance of 80 meters. On October 26, his seventh flight he flew 770 meters in 52.6 seconds and set a new official world record for duration and distance, winning the Archdeacon Cup held by Santos Dumont and a money prize as well This plane was designated a Voisin-50, and was built by the Voisin factory.

Henri Farman went on to win many prizes and started building his own aircraft. He is mentioned again later in this story. Henri became a major aircraft manufacturer in France for many years after his first flight.

Henri and brother Maurice Farman, circa 1908

Pilot 9, Flying Machine 8

A man named Robert Esnault Pelterie became the ninth pilot to fly a powered machine.

Robert Esnault Pelterie
1881-1957

Notice the center wheel under the nose, the wheel on each wingtip, and the 4-bladed prop.

Pelterie made his first flight in this plane on October 22, 1907, and flew 150 meters at Bue, France. His plane had an air-cooled seven-cylinder engine.

Pilot 10, Flying Machine 9

Leon Delagrange
1873-1910

Leon Delagrange became the tenth pilot to fly. His flight took place on November 5, 1907 at Bagatelle, France, at the airport at which the Voison factory was located. His craft, in fact, was built by Voisin. Delagrange became an aggressive pilot in the following year and flew in competition with another early pilot, Henri Farman.

Delagrange's Voisin-built craft

Delagrange's "Airborne I" Machine

Pilot 11, Flying Machine 10

On November 19, 1907, Henri de la Vaulx (1870-1930) flew for the first time. He flew one more time about 70 meters but crashed. Although he was not hurt seriously, he did not pursue aviation thereafter.

Left to right:Gabriel Voisin, Henri Farman,
Henri de la Vaulx

Henri's craft above is believed to be modeled after a Voisin type 2.

Pilot 12, Flying Machine 11

Alfred De Pischoff
1882 – 1922

Alfred De Pischoff constructed his own flying machine using a 25-hp Anzani air-cooled engine. His first flight was on December 5, 1907. He never became prominent as a pilot, even after making roughly six flights, cementing his place in the history of early aviation alongside his French compatriots.

de Pischoff's flying machine

The Wright Brothers' Flyers

After Orville and Wilbur Wright made their first flights in 1903, no one else flew until 1906, and all but one of the flights in 1906 and 1907 were made in France. The exception was one flight in Denmark.

The series of pioneer pilots has been kept in sequence according to the dates of their respective flights. Note that the models flown by these pilots between 1903 and 1908 were not designed by the Wright brothers.

During the next several years after their first successful flights, the Wright brothers continued to modify and improve their original flyer. Each new model looked somewhat different from the earlier ones. These new flyers are discussed in the next chapter.

CHAPTER THREE

The AEA And New Developments

The Aerial Experiment Association

Alexander Graham Bell, the inventor of the telephone, founded an organization called the Aerial Experiment Association (AEA) in 1907. Those that joined together in the AEA are shown below.

Left to right, Glenn Curtiss, Thomas Selfridge, Bell, John McCurdy and Frederick Baldwin

Chairman Alexander Graham Bell
1847 - 1923

Left: Manufacturer Glenn H. Curtiss 1878-1930
Right: West Point Grad Thomas Selfridge 1882-1908

Left: John McCurdy, Canadian Army 1886-1961
Right: Frederick "Casey" Baldwin 1882-1948

The AEA Charter

The charter was made and signed by all five members while they were staying at Bell's mansion near his laboratory in Beinn Bhreagh near Beddeck, Nova Scotia. The document was notarized by a William L. Payzant, Notary Public, on September 13, 1907, at Halifax, Nova Scotia. A copy of the charter is shown in appendix B.

Prior to the Army's involvement with the Wright brothers, the War Department as distinguished from the U.S. Army directly, heard about the AEA, probably because it was formed by Dr. Bell, who was a current member of the Smithsonian Institute board of directors. The War Department wanted to know if there was any interest in the use of a flying machine in wartime.

The War Department sent an army officer, Lt. Thomas Selfridge, to observe the AEA's findings on flying machines. Selfridge was a young West Point graduate who had proven his leadership when he took command of troops sent to help restore order after the San Francisco earthquake.

He was also knowledgeable in the field of engineering. He became one of the four members under the directorship of Bell who built flying machines. His machine was the first one flown by the AEA. Lt. Selfridge was the first Army pilot to fly but not in a military aircraft.

All the AEA members, with the exception of Glenn Curtiss, remained at Bell's mansion until mid-December. Glenn had gone back to assist his trusted associate at his factory in upstate New York.

Projects of the AEA

Dr. Bell had been working on a large box kite that he named the *Cygnet* (French for swan). The *Cygnet* incorporated 3,393 silken tetrahedral cells, stood about 11 feet high with a span of 42 ½ feet. It weighed 208 pounds unmanned. A seat was built in for a man to be aboard the kite.

The U.S. Army officer Selfridge agreed to be the first to fly in it. The *Cygnet* was placed on a raft, which was towed by a steamboat. A tether was also attached to the steamboat. When the boat got up enough speed, the kite rose to a height of 168 feet and stayed aloft for 7 minutes. As the steamboat slowed, the kite floated gently down to the water.

Lt. Selfridge did not get the tow line disconnected in time, however, and the kite was wrecked beyond repair, while Selfridge got really cold in that winter water. Dr. Bell was thrilled that the kite performed so well. At least one member of the AEA had gotten airborne.

Cygnet's Manned Flight
Dr. Bell (right) watches from towboat as the Cygnet leaves the raft and becomes airborne over Baddeck Bay in Nova Scotia.

Glenn Curtiss invited all the members of the AEA to move down to his factory site in upstate New York, where he had a field that could be used as an airport. They all moved down to Hammondsport, New York, and conducted most of their work at this location.

Once the entire membership of the AEA were in Hammondsport, the Curtiss Co. completed the engine for the balloon that Thomas Baldwin had ordered, and he also came to the town to make some test flights. Thomas invited all the members of the AEA to take a flight with him. All members, with the exception of the chairman, actually had a flight in the motorized balloon.

Dr. Bell decided that the planes would be called Aerodromes or just Dromes for short. He also chose to pay each man a salary and it's known that he paid Glenn Curtiss a salary of $5,000 for the first year.

The Baldwin Mix-up

Thomas Baldwin* of California had constructed a balloon and wanted to add an engine to fly and control it. The Glenn Curtiss Company was the most well known manufacturer of special engines, and so Baldwin shipped his balloon to the Curtiss factory in Hammondsport, New York. This company was the only one at that time that

---

* During research on the AEA, I discovered that there were two men with the last name of Baldwin. One man, who was from California, was named Thomas Baldwin (said to be born in 1854), and the other man was Frederick (Casey) Baldwin, of Toronto, Canada. This apparently has caused a mix-up among different authors.

would make a small engine with limited horsepower for a reasonable price. Thomas Baldwin did not become a member of the AEA, but he provided inspiration to continue their quest for a flying machine

French Flyers in 1908

In January of 1908, Leon Delagrange (pictured earlier in this book) received his second flying machine built by the Voisin brothers' company. It was named the *Delagrange II*, and he was airborne in his new version of his first craft on January 20, 1908. This craft had a more powerful engine made by Antoinette, which had 50 horsepower.

In March, Delagrange and Henri Farman were competing at an airplane meet near Paris, and Leon won the prize for making the largest circle of about 700 meters. The French had not mastered the coordinated turns as had the Wright brothers, but they could make some rough circles while airborne.

Henri Farman Racing for Distance

Leon Delagrange Racing for Distance

Henri, the eighth pilot to fly, and Leon, the tenth pilot to fly, met at a field in Issy, France and tried for several prizes: for distance, for best circle, and others. Other early flying machines were there, but they weren't up to the kind of flights that Henri and Leon were, so they aren't mentioned in the stories I have found in my research.

The aircraft meet took place beginning on March 14 through March 17. Both of their craft look similar, probably because they were both built by the Voisin brothers' aircraft company. There were just a few differences between them. The main difference was that Leon's craft had two seats, while Henri's craft only had one seat.

Leon won the prize for the longest and fastest flight, so on the last day of that competition, Henri Farman climbed into the passenger seat beside Leon Delagrange and became the first passenger to ride in a flying machine in France. After the competition was over, Leon Delagrange took his flying machine to Rome, demonstrating his flying ability to the Italians in April and May of 1908. He also took his plane to Turin, Italy for more demonstrations. On July 8 he took up Mme. Therese

Peltier. She later made solo flights which are touched on in later text.

One more Frenchman was to become pilot number 13 to fly. On February 8, 1908, an aircraft mechanic, mentioned only as Boyer, flew for a company named Gastambide-Mengin a tractor monoplane with an Antoinette 50-hoursepower engine, which had a tail at the aft end of the fuselage, but no elevator. No illustration is available.

The Gastambide-Mengin in 1908

Pilots of Record

Dr. William W. Christmas flew on February 8, 1908, at Fairfax, Virginia. Dr. Christmas owned the Christmas Aeroplane Co. and built at least one aircraft. This airplane was the first to use ailerons instead of wing warping, and Dr. Christmas was issued the first patent on ailerons and subsequently sold the patent to the U. S. Government for $100,000. See the epilog for more about the patent. Dr. Christmas was pilot 14.

Dr. William W. Christmas

Three members of the AEA distinguished themselves. Pilot 15 was, Casey Baldwin; pilot 16 was Glenn Curtiss; and pilot 17 was, Lt. Selfridge. All flew on March 12, 1908.

The AEA was ready to build the first drone. Selfridge favored the construction of an unpowered hang glider similar to those built and flown by Octave Chanute before the turn of the century. In fact he wrote to the Wrights seeking advice on glider construction and was answered favorably. The glider was built and flown from a hill in Nova Scotia before the group left for upstate New York.

They took the glider with them and it became *Drone 1*. The glider was fitted with an engine from the Curtiss factory and turned into a man-carrying flying machine. Selfridge was the first member to complete a flying machine, which he named the *Red Wing* because of the red silk that covered the wings.

The *Red Wing* was first flown by Casey Baldwin on March 12, 1908, at Hammondsport, New York. Glenn Curtiss flew it next for 312 feet at an altitude of 200 feet, and finally Lt. Selfridge flew only 120 feet and crashed. The plane was beyond repair.

The Red Wing

The next plane, built by Casey Baldwin, was named the *White Wing* (white cloth-covered wings) was flown by Baldwin, Selfridge, Curtiss, and McCurdy later in March the same year.

The White Wing

1. Casey Baldwin   279 ft. May 18, 1908
2. Tom Selfridge   240 ft.
3. Glenn Curtiss   1017 ft..
4. Doug McCurdy   549 ft.

Before listing the next group of pilots, it's interesting to note how records were kept on these first pilots: After McCurdy became pilot number 18, the 21st century researcher, Earnest Jones, finished off his records at the end of 1908 with a total of 35 first pilots in sequential order by date. All the rest, with the exception of one Englishman and one German, were French. In this author's research, he noted a few pilots' first flight numbers were out of sequence. Two members of the AEA, Lt. Selfridge and Glenn Curtiss, were numbered in reverse sequence, because on March 12, Selfridge was the third pilot to fly, but was the one who ended in a crash – not Glenn Curtiss. The author also discovered that was not an Englishman, but rather a German. Hans is discussed later when he flew in 1908.

Pilot 18, May 18, 1908.

Douglass McCurdy became the 18th man to fly a powered flying machine.

Pilot 19, July 22, 1908

Captain L. F. Ferber of France flew in the *Ferber IX*, a reproduction of a model of 1905. Ferber was one of the pioneers in flight experiment, interrupted by other work. He began in 1898 and collaborated with Chanute by correspondence, who converted him to a plane with two wings, a biplane.

Pilot 20, August 4, 1908

Paul Zens of France flew a pusher biplane with elevator and rudder forward. Using an Antoinette 50-hp.engine, his plane sort of jumped off the ground.

Pilot 21, August 19, 1908

Georges Legagneux of France flew a *Ferber IX* (Antoinette III) for a distance of 256 meters and won the third prize of the ACF.(Aircraft Club of France). He was a mechanic for the Antoinette Company.

Pilot 22, August 20, 1908

Welferinger of the Antoinette works flew the Gastambide-Mengin machine up to 100 meters at Issy, France. He carried a passenger, Robert Gastambide. *This was the first monoplane to carry two persons for 100 meters*

Pilot 23, September 9, 1908

Balloonist René Gasnier of France made several flights in a pusher biplane.

Pilot 24, September 29, 1908, in England

S. F. Cody, an American living in England while working in aeronautics for the British Government, made his first flight of about 78 yards. The plane was a twin screw tractor biplane.

Pilot 25, October 6, 1908

Gabriel Voisin, discussed along with his brother Charles in Chapter 1, was the younger brother of Charles (12 years younger), and when Charles was killed in an auto accident, Gabriel became the head of the Voisin Aircraft Manufacturing Company. He sold the company shortly after World War I.   Their 1908 plane is also pictured in Chapter 1.

Next pilot, October 28, 1908

Hans Grade (1847-1946

Pilot Hans Grade was listed as an Englishman, but in fact, he was actually the first German pilot and should have been listed as the 28th pilot.

Hans Grade was an engineer and a German aviation-pioneer. On the October 28, 1908, he successfully conducted the first motor -flight over German soil in a motorized aircraft of his own construction at Magedburg.

A year later, on the October 30, 1909, he won the 40.000 Reichsmark "Lanz-Preis der Luft," as he was the first German to fly a flat "8" in a German aircraft with German engine around two pylons, standing at a distance of 1,000 meters (no match for pilots from other nations at that time).

Hans Grade's 1908 Monoplane at Dresden Airport

AEA Flights Early in 1908

Although Glenn Curtis had started building his plane in 1904, he had a business to run and had less spare time than other members of the AEA, so he had his plane ready about the same time as the other three members who were building planes. Dr. Bell thought the plane looked like those red June bugs that come out in the summer time. Curtiss took his suggestion and named his plane *June Bug*,

The June Bug Before Wheels

Curtiss first flew the plane he designed on June 21, 1908. He flew a distance of more than 3,000 feet. Sometime in the early spring of 1908, Aero Club America, formed earlier, started to promote flying machine development (the Wright brothers had joined the club). The club joined with *Scientific American Magazine* to give a prize to the first flying machine that could take off on its own and fly a distance of 1 mile. A silver trophy was displayed in New York in a store window to advertise what was called the *Scientific American Trophy.*

The Aero Club asked the Wright brothers if they would like to enter the contest, but they refused because they had not yet perfected a wheel system for their machines and were still launching them from a platform.

After Glenn Curtiss made his June flight, the AEA contacted the *Scientific American* and told them that they were ready to try for the prize. The Aero Club and the *Scientific American* decided that there should be a public meeting on July 4th in Hammondsport at Stony Brook Farm.

On Independence Day in 1908 people came by train, car, and buggy to the farm where the event was to occur. It turned out to be rainy and windy in the morning, and so the flying attempt was put off. There was a winery nearby, so they opened their doors, and the sponsor furnished a buffet while the winery furnished wine. People were having a good time. About 5 p.m. the weather seemed to clear, and so the *June Bug* was wheeled out of its hangar on its newly installed wheels and pushed to the starting point of the contest.

The plane was to take off and fly for one full mile before landing. A red banner was laid out exactly 1 mile from where the plane was to get airborne and so the plane would have to land just beyond the 1-mile marker.

Sometime after six 'o clock in the evening, Curtiss got airborne and was planning on flying the full course; however, his plane kept climbing and was up to 40 feet high when it should not have been over about 20 feet high. Glenn cut his engine and landed. His plane was again rolled back to the starting point, and the tail was adjusted so the plane would not climb so much, and Curtiss took off again. At 7:30 p.m. on that day of July 4th he flew 20 feet above the earth, disappeared over the horizon, and landed just beyond the red banner that marked the 1-mile distance flown.

There were several observers, one of whom was the pilot of Langley's plane that crashed into the Potomac River in 1903, Charles Manley. Manley was a graduate engineer who had worked on many Langley flying machines. Manley verified that Glenn Curtis had landed beyond the red banner, so there was no question about it.

Glenn Curtiss' Flight at Stony Brook Farm

The Scientific American Trophy

The Wright brothers were glad to see Glenn Curtiss win the Silver Cup but not too happy that he received a cash prize as well. They asked him to pay a patent fee. Curtiss ignored them, setting the scene for a long lawsuit about the first flying machine patent received by the Wright brothers. The lawsuit was not begun immediately, but when Glenn won another prize in 1909 that included cash the Wrights did sue. Details of this fascinating lawsuit are provided later in this text.

Curtiss' Red Wing in 1908

The previous photo clearly shows that Glenn Curtiss did not use wing warping to make smooth turns but used ailerons instead. If you look closely at both wing tips of the *June Bug* photo, the plane without wheels previously shown, you can see the actual ailerons on the right and left wing tips. Dr. Bell applied for a patent on several improvements on items developed differently from those used by the Wright brothers, but in the case of the ailerons, he used a different designation.

That may be the reason that an actual patent for the aileron was given to Dr. William W. Christmas, who is pictured on page 31. His ailerons were similar to those used on the light aircraft of today, not so much like the wing tip ailerons on the *June Bug* shown earlier. See the epilog for a discussion of ailerons with photographs of the Christmas aircraft.

Wright Brothers' Bid

Late in December of 1907 Wilbur Wright met with an official from the U. S. Signal Corps and Board of Ordnance to discuss their airplane's capabilities. The Signal Corps then advertised for bids for a military heavier-than-air flying machine to be submitted by February 1, 1908.

On January 29' 1908, the Wrights submitted their bid to the U.S. Signal Corps to supply a heavier-than-air flying machine, weighing between 1,100 and 1,250 pounds carrying two passengers, and flying at a speed of 40 miles per hour. Their bid price for this plane was $25,000.

Surprise Bid from the Ghost

About that same time a surprise bid was made by Augustus Herring and it was the lowest bid of $20,000. His plan was to obtain the award by making the lowest bid and then subcontracting the building of the machine to the Wright brothers. His plan was foiled when the Army accepted the bids by both Herring and the Wrights.

On February 8, 1908, the bid to furnish a flying machine to the U.S. War Department for a sum of $25,000 was accepted. At the same time the War Department accepted the bid of $20,000 by Herring.

Herring, in an attempt to save face, said he would provide an airplane and fly it to Washington. After the Army had given him numerous extensions on the due date of September 28, Herring stopped the charade by formally requesting his contract be voided for reasons of nondelivery because, he said, he had sold his planes to another country.

The Wrights Prepare for September 1908

Wilbur and Orville again took a plane to Kitty Hawk, later to be called the *Model A*, and Wilbur took up a passenger on May 14.

1907 – 1909 Wright Model A

This was the aircraft that convinced the world that the Wrights had indeed flown. It was also the first two-seat aircraft, and the first Wright aircraft in which the occupants sat upright.

Wilbur in France

Wilbur Wright had the *Model A* shipped to France on the same ship on which he traveled, leaving the port of New York in May 1908. Orville was left in Dayton to build another *Model A*, just like the first, to demonstrate to the U. S. Army in late August of 1908. Arrangements had been made with Frenchmen whom Wilbur had met earlier, so the demonstrations were to be at the Hunaudiéres Race Course about 5 miles south of Le Mans, France.

Wilbur arrived in France on May 29, and as soon as he could get his craft cleared through customs, he traveled by train with his *Model A* to the racetrack. With the help of friends he unpacked his plane and started to assemble the craft. A small hangar had been provided.

When he opened the crates containing the unassembled *Model A*, Wilbur, who usually had an even temper became very angry. The machine had been badly damaged in transit:. The radiators were smashed, seats broken, axles bent, and ribs cracked. He sent a blistering letter to Orville: "I opened the boxes yesterday, and have been puzzled ever since to know how you could have wasted two whole days packing them…I am sure that with a scoop shovel I could have put things in within two or three minutes and made fully as good a job of it."

As it happened, Orville had not been to blame. French Customs had opened the crates and then had poorly repacked them, causing the damage. But that did nothing to relieve the pressure on Wilbur, with the newspapers taunting the Wright brothers as bluffers and not real flyers. Wilbur was once again in need of a real friend, and one appeared in the person of Leon Bollee, an automobile manufacturer and a sport balloonist, who gave Wilbur a place to work near the racetrack at Le Mans and provided a team of mechanics to help him with repairs.

*Small Work Hangar
at Le Mans Racetrack*

*Flyer A Before Repair
at Le Mans Racetrack*

Although the mechanics under Wilbur's instructions were a great help, he himself had to do most of the detailed tasks, such as sewing and stretching wires until his hands were raw. He even burned himself when a radiator hose tore loose from the engine and sprayed him with boiling water. It took 6 weeks to put the airframe together, twice as long as had been planned. Then it rained for several more weeks before he could fly.

The weather finally cleared, and on Saturday, August 8, the day arrived that Wilbur had been waiting for almost three years. The sky was clear, the wind was nearly still, and his flyer was waiting. Later he wrote to Orville, "I thought it would be a good thing to do a little something."

A small crowd had gathered in the bleachers just in case Wilbur decided to fly without giving notice. One of those in the crowd was the Frenchman who had flown about a year earlier in his own machine, Louis Bleriot, and he was anxious to find out if the Wright brothers really had a genuine flyable machine.

Flyer A, almost ready to fly August 8, 1908

Wilbur and His Flyer with Two Seats in France

On August 8, 1908, Wilbur Wright made his first demonstration flight in the *Model A* in Europe at the racetrack near Le Mans, France.

Wilbur got his engine going and took off with the aid of a man on his right wingtip to keep the machine steady until it reached flying speed. The *Model A* still had no wheels and took off from a sloping ramp. This time the helper on the wing had only to steady the plane for a few feet before it was airborne and on its way.

Wilbur, flying alone with one seat empty, made two figures 8 above the field, and there was no longer any question as to whether his plane would fly and make banking turns just like a hawk. France no longer questioned the Wright brothers' ability to fly a heavier-than-air machine.

Louis Bleriot told some reporters who were not at the field that day that no French pilot, including himself, could fly a plane in perfect circles that came anywhere near to the circles that Wilbur Wright had made. He was quoted as saying, "I would have waited ten times as long to see what I have seen today. Monsieur Wright has us all in his hands." The news was out, and the next crowd at the racetrack was tremendous.

Another French pilot, Henri Farman, who later established an aircraft factory, was also there. Two days later on Monday, August 10, crowds estimated at 2,000 watched Wilbur on his first flight of the day make a three-quarter circle and land. He then flew again for about 2 minutes, making two large figures 8 over the field at a height of 30 to 40 feet above the ground.

The next day before an estimated crowd of 3,000 spectators, Wilbur flew one flight for about 4 minutes at heights from 2 to 20 meters (3 to 70 feet), making three large circles around the entire field. The field was about 800 meters long and 300 meters wide.

Wilbur made more demonstration flights on August 11, and on August 12 he made three flights. The three flights on August 12 lasted 15 minutes, 20 minutes, and finally 29 minutes. On his last landing, however, there was a slight mishap causing repairs to be made to one spar, several ribs, and one skid. On the whole, however, the meet was highly successful for the Wright brothers.

CHAPTER FOUR

Demonstration Flights & Contracts

After the minor accident with his flyer, Wilbur Wright knew that he would need time to rebuild that damaged part of his craft. The race course had been advertising that the races would start just a few days after Wilbur had his slight accident. The group that wanted to see more flying arranged to help Wilbur move to a new field only a few miles away

Moving To A New Field

Wilbur had his plane moved to another field about 8 miles away because the racing session was to start on August 14[th.] After he got settled at the new location, he made his repairs and flew demonstration flights for two more days.

Wilbur takes an evening flight over France in 1908 in his Model A

Once Wilbur finished his demonstration flights in France, a French business group signed a contract with Wilbur that would finally bring some money to the brothers. Their contract called for them to furnish four two-place machines at $4,000 each, make two demonstration flights of 31 miles, each carrying a passenger, and teach three men to fly and solo.

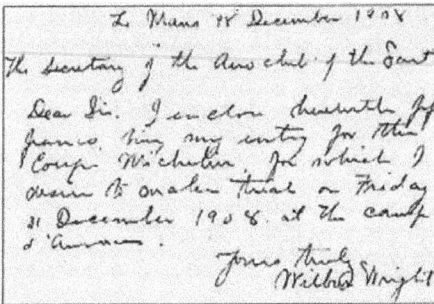

Wilbur's Letter to the U.S. Aero Club from Le Mans
(1909-02—07 CAvM, Austro-Hungarian Collection, L'Aérophile, 15 January 1909.)

The year's most famous flight occurred on December 31st and lasted 2 hours, 20 minutes. As he landed, Wilbur knew he had won the 1908 Michelin cup, a closed-circuit distance competition he had entered on December 28th. The short note above is a facsimile of Wilbur's entry.

After Wilbur finished his demonstration flying in France, he also flew in Italy and in Germany, but the exact dates and records are somewhat sketchy.

The following statement was obtained from a joint publication of the U.S. Centennial of Flight Commission and the National Aeronautics and Space Administration, published in 2003:

> The Wright brothers' scrapbooks and contemporary accounts in American and European newspapers and in aviation and technical journals were the primary sources for the remaining Wright flight logs. No diaries or notebooks were maintained by Wilbur in France in 1908 recording his flights at the Hunaudières Race Course and Camp d'Auvours at Le Mans, or in 1909 at Pont-Long, at Pau, and at Centocelle Flying Field, Rome. Nor did he keep detailed records of his flights at Governors Island, N.Y., and at College Park, Md., in 1909.

The last sentence in the preceding paragraph becomes relevant when Wilbur Wright makes his famous flight from Governors Island, New York, for the 1909 historic fair celebration. This was a grand affair in recognition of Henry Hudson's discovery of the Hudson River and the steamboat trip up and down the Hudson by Robert Fulton many years ago.

Below is a copy of the cover of a French magazine in which a great story was published on Wilbur Wright and his demonstration flights at Le Mans in 1908.

Cover of Booklet Published by
Distinguished Author François Peyrey

The updated edition of *Les Premiers Hommes-Oiseaux*: Wilbur et Orville Wright, signed by author François Peyrey, was published in January 1909. This respected aviation journalist included a detailed list of all flights made by Wilbur near Le Mans in 1908.

(1909-02-06 CAvM, Austin-Hungarian Collection, Les Premiers Hommes-Oiseaux: Wilbur et Orville Wright front cover)

First Helicopter Flight

In November of 1908 the first helicopter was flown by Paul Comu, a French inventor. The flight lasted only 20 seconds and hovered just 1 foot (30 cm) above the ground.

Wilbur's Last Flight of 1908 in Le Mans

On the last day of the month December 31, 1908, Wilbur stayed aloft for 2 hours and 18 minutes to win the 20,000-franc Coupe Michelin prize. The nearest competitor was Henri Farman, who had stayed aloft for 45 minutes back in October.

Fort Myer, Virginia

Orville's Demo for the Army
Ft Myer, Virginia, USA (the area in brownish yellow)

Wilbur's Demo Flight
He flew from a racetrack south of
Le Mans near area of red circle

Orville's Demonstration Flights for the Army Signal Corps The photo below is of the military flyer (*Model A*) that Orville flew at Fort Myer in 1908 Small wheels had been added as required by the Army.

Orville was flying in America at this same month for the U. S. Army demonstration at Fort Myer, Virginia. On August 27, 1908, Orville Wright had reassembled the military flyer (*Model A*) at Fort Myer, the U. S. Army's weapons testing base.

Orville Wright in Pilot's Seat at Fort Myer

Military Flyer Airborne Over Crowd at Fort Myer

The Army was not impressed when Orville took Army Lt. F.P. Lahm on a short flight. On September 9, however, Orville took a passenger up and circled over the parade ground for more than an hour. This impressive feat was viewed by Army personnel as well as a large crowd of invited civilians. The next few days he took many passengers on demonstration flights.

Tragic Event

On September 12th, Lt. Selfridge, discussed earlier in the book and a member of the AEA, was to ride with Orville. As Orville was making climbing circles over the crowd, two loud thumps were heard, and as Orville reached to cut off the engine, the airplane gave a violent shake.

Selfridge, left, Orville, right
Ready for Takeoff

One propeller had splintered and caught the aircraft rigging, causing the craft to plunge to the ground, smashing nose first, burying the pilot and his passenger in a twisted mess of wood, wire, and cloth.

Photo of Orville's Crash with Selfridge

Orville Being Carried Away on a Stretcher

Orville was sent directly to the military hospital at Fort Myer.

Body of Lt. Selfridge after Removal from Crash

Lt. Selfridge, graduate of West Point, died in the crash on September 12, 1908, becoming the first person to be killed in an airplane crash. Orville, seriously injured, suffered a broken leg, broken ribs, and an injured back, but he lived to fly again.

The family was notified immediately about the crash, and Katherine took indefinite leave from her teaching job in Dayton and rushed quickly to her brother's side at the hospital in Fort Myer.

Katherine Wright (1874-1929)

Wright Family Home Dayton,
Ohio, Built around 1872

When Wilbur, who was in France, learned of Orville's injury, he realized that he should not have let Orville do the demonstrating to the Army all alone and considered returning to the United States.

Orville dictated a letter to his sister, Kate, that there was no need for Wilbur to come home but rather to complete his European mission. Orville told Kate to tell him that he would be O.K. but he would remain in the hospital for a while and that the Army had agreed to postpone the flying trials till later. Both Orville and the Army agreed that the accident was not anyone's fault and that accidents sometimes happened with modern machines.

Wilbur then decided to stay in Europe and continue demonstrating his *Model A*. He flew in Italy and Germany before he was joined in 1909 by Orville and their sister.

Wilbur and Plane in Germany - 1908

Below is the cover of a French magazine that shows photos of the fourteen French and two American flyers, Wilbur and Orville, who had flown in France by the end of 1908.

Significant Flight by Wilbur

Charles S. Rolls (1877 - 1910)

On October 8 Wilbur was still flying demonstration flights at Camp d' Auvours, France, and his passenger on this date was none other than Charles S. Rolls, who had become a partner in the Rolls-Royse. His flight from England to France and back is described in the epilog.

Ready for Takeoff
(Charles Rolls on left and Wilbur on the right.)

Wilbur gave Charles a few flying lessons, and Charles ordered a Wright aircraft from the Wright agency that had been established in France. An agency had also been established in Germany by this time.

The US Army's first airplane at Fort Myer, VA, 1909

Four Wright flyers were shipped to the French agency and one was shipped to the German agency during the years of 1909 and 1910. The military version shown above was the same as the three shipped to France and the one shipped to Germany.

Pilots in Sequence Series Continued

Pilot 26, October 29, 1908

Maurice Colliex of France piloted a triplane and made several hops at Issy, France.

This particular flying machine is pictured, because it was the first craft listed as a triplane (3 wings).

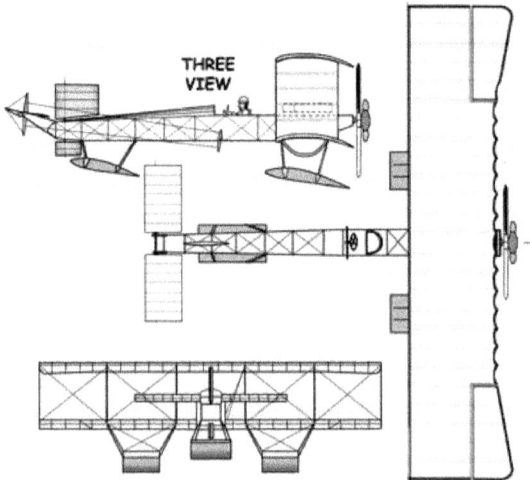

THREE
VIEW

This plane was built by the Voisin factory, and the lower wings were influenced by the pilot Bleriot, who was unconventional in many of his designs and worked with the Voisins.

Pilot 27, October 31, 1908

Baron De Caters made his first flight in a Voisin triplane.

Pilot 28 was listed as Hans Grade of England. He was actually from Germany and has been discussed previously as the first German pilot.

Pilot 29, November 3, 1908

An Englishman named Bellamy did get airborne but later crashed. Although he was only slightly injured, he did not fly again.

Pilot 30, November 9, 1908

A French pilot named Goupy flew a Voisin triplane for a short distance at Issy. Issy, France was the airfield where the Voisin brothers had their aircraft factory.

Pilot 31, November 9, 1908

An Italian pilot, Lieutenant Mario Calderara, flying at Issy, France, flew a Voisin plane after Wilbur Wright had given him a few flying lessons at the Centocelle Flying Field a month or so earlier.

Pilot 32, November 19, 1908.

A man named Zipfel, a German, made several flights at Issy, France, and in January of 1909 made exhibition flights in Berlin.

Pilot 33, November 20, 1908

J. C. Moore-Brabazon of England made his first flight in a Voisin plane and flew at Issy, during the month of November 1908, at the location of the Voisin factory.

John T. C. Moore-Brabazon
1884 - 1964

Voisin Plane Flown by Moore-Brabazon

Before this flight John Moore-Brabazon was a companion to Charles Rolls, a partner in the Rolls Royce

Auto Company, and they went up together in balloons several times before they tried flying machines. These men and their adventures are described further in the epilog.

Pilot 34, November 21, 1908

At Bue, France, a Frenchman named Chateau flew the Pelterie II for 316 meters thus winning prize by the A.C.F for flying over 200 meters.

Pilot 35, December 18, 1908

Another Frenchman, Melvin Vaniman, flew his triplane with an Antoinette 70/80 horsepower engine at Gennevillier.

Triplane Built in England by the Avro Company

I was unable to acquire a picture of the French triplane flown by Vaniman, the 35th pilot listed above, but this plane was copied after his aircraft of 1908.

CHAPTER FIVE

Flying Machines 1908 - 1909

Thérèse Peltier

According to my research, a Frenchwoman, Thérèse Peltier, was perhaps the first woman passenger in a heavier-than-air craft although that honor is also given to Elsie Raymonde Derroche (see page 21). At any rate she is a prominent woman pioneer.

Thérèse Peltier, pictured below, was a sculptor who became the first woman passenger in a heavier-than-air craft. On July 8, 1908, she made a flight of 656 feet with her instructor, Leon Delagrange, in Turin, Italy. She later made several solo flights in a Voisin biplane but did not pursue a flying career.

AEA Meeting

Chairman Alexander Graham Bell called for a meeting of the AEA on January 24.

The chairman and all the members with the exception of Glenn Curtiss were in attendance. It was obvious that the meeting was not going well, and so the members, in agreement with Dr. Bell, voted to dissolve the AEA since they had essentially accomplished their original mission. Dr. Bell filed for dissolution of the recorded organization in February 1909.

With Bell's permission, John McCurdy took the AEA's *Silver Dart* to Nova Scotia to show many of his Canadian friends as well as neighbors of the Bells that airplanes were now actually flying through the air.

Curtiss and Herring

The ghost of our story, Augustus Herring, finally met with Glenn Curtiss. He assured Curtiss that he had patents on file for a flying machine and that he had other patents as well. Curtiss believed the smooth-talking conman, and they formed a corporation, the Herring-Curtiss Company, a company to build airplanes. Curtiss put up $300,000, and Herring put up his patents valued a $300,000 (according to Herring). If he really had had those fictitious patents, they probably would have been worth that much because the potential lawsuit by the Wright brothers would not have been effective if filed.

Research by the author indicates that Augustus Herring claimed that he had applied for patents on flying machines before the Wright brothers. All indications are, however, that he had no patents, and there is no record that he had ever even applied for a patent of any kind.

Curtiss Reims Racer he took to France in 1909

First Flight in Canada

*AEA's Silver Dart Flown by John McCurdy
on February 23, 1909, in Nova Scotia, Canada.*

Reims, France

Glenn Curtiss preparing for the First International
Air Meet at Reims, France, to be held August 22 through
August 29, 1909

*Glenn Curtiss Flying at Airshow at Mineola,
Long Island, New York*

Curtiss at Mineola Fairgrounds

Glenn first flew his modified Golden Flyer at his home field in upstate New York on June 26, 1909. He was planning on taking this plane to the big international race at Reims, France, in August 1909. At the fair in Mineola he won two prizes.

The first one was put up locally for a plane that could fly 1 kilometer (about 0.6 mile). He won a $250 prize for that flight on July 17. He next won the second trophy sponsored by the *Scientific American* that had a cash prize of $3,000 and a beautiful trophy for flying 25 kilometers (about 15 miles).

Scientific American Trophy

Wright Brothers' Lawsuit

The brothers filed their promised lawsuit against the Herring-Curtiss Company for violation of the patent law. See the epilog for details on the patent wars.

Germany's First Flying Machine

Igo Etrich's tractor monoplane

A German engineer, Igo Etrich, designed his tractor monoplane that first flew on July 20, 1909. This photo is a replica of the original aircraft flying in Germany. Igo named this plane *Taube 1* (the German word for dove). This plane had a 100-hp water- cooled Argus inline engine.

This aircraft was more advanced than any other aircraft developed at that time. The German engineers had studied the Wright flyer purchased by the German government, as well as many of the aircraft that the French had produced in 1908, and the shape and style of this plane are similar to the first plane to cross the English Channel, which was developed and flown by Louis Bleriot.

German Taube 3

This German museum piece is the World War I replica of the first aircraft used in that war. This plane was a later model of the *Taube 3* shown above. This newer version used a Mercedes engine of over 125 hp. Its wingspan was 46 ft,. and it could cruise at 60 mph.

The *Taube* was first used as an observation plane and was the first stealth aircraft. It was covered with a nearly transparent material that made it very hard to see on a normal day over France.

Crossing the English Channel

Painting of Bleriot

Early in 1909 Bleriot attempted a flight across the channel, but his engine quit just as he started to cross the Channel from France. He was able to recover most of the plane and rebuild it. Then on July 25 he made a successful crossing.

*Bleriot crossed the English Channel in this plane
on July 25, 1909*

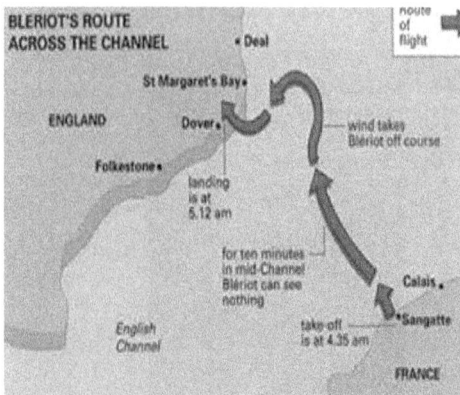

A wind shift occurred as Bleriot approached the
English coast, and visibility was limited. He then turned
back toward Dover. That maneuver explains the big curve
shown in his route before he landed in a low spot near
Dover Castle.

His plane could not climb high enough to land on the
white cliffs of Dover. The Channel crossing time was 47
minutes. See the epilog for a photo of the flight memorial.

U.S. Army's First Flying Machines

It was proven that the crash of the Wright military flyer was not Orville's fault, but rather that one of the two propellers had fractured and cut through a control cable. Orville had already proven that his plane was airworthy, and the Army honored its contract and gave an extension to allow the Wright brothers to complete the contract after Orville recovered from his injuries.

At this time Wilbur had returned with his sister Kate, to the United State while Orville was giving demonstrations of a newer version of their Flyer in France, Italy, and Germany. Orville did return to the U. S in the fall of 1909.

Army Trials

On June 20, 1909, Wilbur Wright returned to Washington with a new and somewhat improved airplane, the 1909 flyer. Official trials began on July 27 when Wilbur flew 1 hour, 12 minutes, 40 seconds with Lt. Lahm on board as an observer.

The final trial flight was made on July 30 when Orville flew the airplane at an average speed of 42 mph with Lt. Foulois as an observer aboard. The 1909 Wright flyer was formally accepted on August 2 and was designated as the Signal Corps Airplane No. 1, thereby becoming the world's first military airplane.

Wilbur Boards Plane for First Flight at Fort. Myer in 1909

This was a more advanced model of the plane that Orville had flown a year earlier. It had a more powerful engine and larger wheels as well as other improvements. The first flight was with Lt. Lahm as a passenger on June 27, 1909.

During his demonstration this year, he took up several of the officers who were on the aeronautical board. A photo of the entire aeronautical board consisting of seven military officers is shown on the next page.

Photo of Lt. Lahm and Wilbur Wright Ready for takeoff at Fort. Myer on July 27, 1909

*Lt. Foulois meeting with the Wright brothers at Ft. Myer
in the summer of 1909*

This was the first of the two sessions that Wilbur held with the Army personnel in 1909.

*Left to Right: Lt. Frank Lahm, Lt George Sweet,
Maj. Charles Saltzman, Maj. George Squier,
Capt. Charles Chandler, Lt. Benjamin Foulois,
Lt. Frederick Humphreys*

All phases of the required demonstration flights were completed and accepted by the board with the exception of the requirement that two military pilots receive flight instructions. Those two pilots were selected to be Lt. Lahm and Lt. Humphreys.

The War Department decided, upon complaints of the other Army branches, that they had work to do of their own at Fort Myer and that the flight training must be moved to another field. The board made a joint decision with Wilbur's approval that he had completed this phase and when they found a suitable field they would then have him train the two pilots wherever the field would be located.

(Please note that Lt. Foulis was not selected for pilot training at this time. He did have a flight as a passenger however. The follow up story, later in this same year, explains how Foulois still managed to become the first authorized pilot of the Army Air Corps.)

Katherine on Flight with Wilbur

This flight was made at Hoffman Prairie, Dayton, Ohio, on August 7, 1909, in preparation for flying at the Henry Hudson – Robert Fulton celebration in New York. A week later, Katherine and Orville left the United States for a flying exhibition in Berlin to satisfy two additional contracts.

Orville in Germany

The above photo is of Orville in Berlin, in the late summer of 1909, talking with Prince Fredrick Wilhelm. The man in the derby is Hart O. Berg, German agent for the Wright brothers in Germany.

Orville made a few flights with passengers for a few days before leaving Germany

First International Airshow Riems, France

Now we move to the great first international air meet to be held beginning on August 22, 1909, at a specially built field at Reims, France. It was similar to the World Olympics in the middle to late 20th century a rail line was built running from downtown Reims. The Wright brothers were already overbooked back in the United States, and

Orville, still in Europe, was about ready to come home. Neither of them went to the big event in Riems. At this time you could get from Europe to a port in the United States in only 5 or 6 days.

Curtiss Flying Machine in Front of the Reims Grandstand

Not only was the grandstand filled but thousands of observers also stood around the specially built field because everyone wanted to see flying machines in the air. Estimates were that 5,000 were in the grandstand, and about 500,000 were gathered about the perimeter.

*Four flying machines in the air at the same time
at the first international air show. It was the first airport
built especially for such an event at Reims, France.
Location of the Big Airshow of 1909*

Reims is North East of Paris

Glenn Curtiss at Reims Airshow of 1909

There were many prizes and there were two major money winner flights. One major prize was for the flight that went the greatest distance. The other major prize was for the fastest planes flying around three pylons for three laps. There were many other flyers and there were other smaller prizes for different flying trials such as highest altitude, tightest circle, and acrobatic demonstration for 5 days.

Henri Farman raceing for distance
prize in France 1908

Henry Farman

Curtiss Rounding the Last Pylon

Glenn Curtis won the first international speed event by flying at a speed of 75.6 km/h (47 mph).

The Wright Brothers and the Airshow

Eugene Lefebvre

As chief pilot for the French Wright Company, Eugene Lefebvre was the only stunt pilot at the Reims airshow. His stunt was to dive at the ground and pull out just before it was too late. Only 2 weeks after Lefebvre finished his stunt at the airshow, he became the first French pilot killed in a flying machine when he crashed while trying to improve his stunt.

Settlement of the Wrights' Lawsuit Against Curtiss

On September 30, 1909, Judge John R. Hazel, U. S. Circuit Court, ordered the Herring- Curtiss Company and Glenn Curtiss to stop manufacturing airplanes. This of course was a ruling in favor of the Wright brothers. See the epilog for the details of the patent wars.

Wright Brothers' Incorporation

The brothers had been operating as a partnership, and to stay in the business they needed more capital with which to buy needed equipment.

Some major investors, having agreed that airplanes were here to stay, suggested that they would supply them with more capital if the brothers would agree to setting up a corporation so that shares of stock could be sold. Two of the best known investors were

J. P. Morgan and Cornelius Vanderbilt. All the investors understood the stock market and how it operated and eagerly joined in to buy the initial stock offering.

Morgan and Vanderbilt negotiated a contract with the Wright brothers that would pay them $100,000 cash and one-third of the shares of stock. The initial stock offering was $1 million, and Wilbur was to be the president of the new Wright Airplane Corporation. This corporation was registered in the state of New York on November 22, 1909. The Wright brothers were now fully in the business of manufacturing airplanes.

Historic Fair Celebrating Henry Hudson and Robert Fulton

In 1905 the New York State Legislature appointed a commission to organize a naval exhibition, "great land parades," pageantry, and festivals stretching from Staten Island to Troy and Cohoes, New York, 200 miles north. It was to take place in the fall of 1909 to mark the progress of American civilization since the explorer Henry Hudson's voyage of discovery as he sailed up the Hudson (then North) River in 1609, and the inventor Robert Fulton's navigation of the river by steam power in 1807.

It is said that careful pains were taken to avoid anything of a commercial nature.

With the development of the flying machine, the existence of such machines became widely acknowledged, and yet most of the general public had never personally witnessed a flight. Therefore, the State of New York invited both the Wrights and Curtiss to fly up, down, and around Manhattan Island and along the Hudson River. They accepted this invitation to demonstrate their respective planes and prove to all that man really could fly through the air in a flying machine.

When the exhibition began in late September of 1909, the commission advised them that they could use Governor Island as a flying field since it was in the bay and had a large flat area that would accommodate launching and landing of aircraft. Both Wilbur and Glenn were planning on flying beginning on September 29; however it was at first foggy in the morning and excessively windy later in the day, so neither of them were able to fly that day.

(There is some mystery that Glenn might have flown in the light fog, but this has never been proven.) This mystery arose because Glenn had another commitment to fly at a fair elsewhere on November 1, so he packed up and left the next day and did his demonstration flight in that area about a year later in 1910.

On September 30, 1909, Wilbur Wright flew his newest flyer, *Model B,* over the harbor and up and down the Hudson, flying under the George Washington Bridge and making a circuit around the Statue of Liberty. Thousands of New Yorkers and even New Jersey citizens saw a real flying machine carrying a man over Manhattan and up and down the Hudson Bay.

*Wilbur Wright rounding the
Statue of Liberty, September 30, 1909*

Multiple Passengers

On November 3, 1909, Dr. William H. Green, a dentist, flew his biplane with three passengers aboard at Morris Park, New York. No photo is available.

New Army Field for Pilot Instruction

Following the acceptance of the 1909 flyer, aviation activities were moved to College Park, Maryland, where a large field was available. On October 8, Wilbur Wright began giving flying lessons to Lieutenants Lahm and Humphreys. Humphreys soloed on October 26, thus becoming the Army's first trained pilot. Lahm soloed several minutes later. Within several weeks, both officers were ordered to return to duty with their respective Army units. The Aeronautical Division thus was left with one airplane and a handful of airplane mechanics, but no pilot.

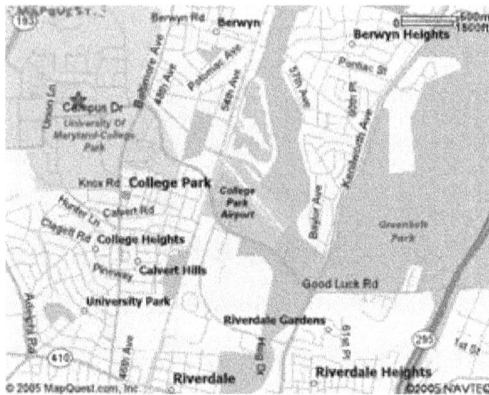

In 1909 the U. S. Army leased the land (299 acres) where the College Park Airport is today. (See map above.) They moved the Wright flight evaluation from Ft. Myer after they approved the Wright brothers' military flyer, a two-seated aircraft.

Wilbur trained the first two army pilots at this field and left the flyer with the Army. In the center of this map you can see the College Park Airport. Although the Army kept this field for a few years, they closed it and turned it over to the owner of the land, and eventually it became a civilian airport. A short discussion of the airport's military history is explained in the epilog.

Before the end of 1909, the Army's evaluation board, with coordination or orders from Army headquarters, appointed Lt. Benjamin D. Foulois as the first member of the Army Air Corps, and he was issued orders to transfer to Fort Sam Houston in San Antonio, Texas. One airplane was sent to Sam Houston along with three Signal Corps enlisted men to act as mechanics for Foulois' aircraft. One factor was that the weather in San Antonio was much better for flying than that at the airport at College Park.

At that time Lt. Foulois was the only man who was actually in the Army Air Corps and in fact was the entire Army Air Corps. The enlisted men with him still belonged to the Signal Corps and were on temporary orders to Fort Sam Houston. Known as "the father of military aviation," Lt. Foulois' experience in Fort Sam Houston is covered in more detail in the epilog.

The other two other planes were shipped from the Wright factory direct to North Island near San Diego along with some other men to form a school to train mechanics and others, still under the Signal Corps department.

Glenn Curtiss' Return to Hammondsport

Curtiss returned from Reims with $15,000 in prize money, feeling very good about his accomplishments. When he met Herring at the Herring-Curtiss Airplane Corporation and checked the books, however, he must have used a curse word or two. The company had been badly mismanaged and was on the brink of collapse. Of course the Wright brothers' court order forbidding the manufacture or sale of airplanes was also a factor. Without Curtiss' knowledge, Herring had entered into contracts to sell airplanes through two different department stores, one in Chicago and another in New York.

Ad Placed in a New York Paper by Herring

Curtiss and his board of directors secretly agreed that Herring had to go, but how?

The eventual solution was bankruptcy. Curtiss also moved his motorcycle manufacturing machinery to another plant in town, and renamed his plant Marvel. Details about this subject continue well into the future. See the epilog about events after 1909.

EPILOGUE

Events Following 1909

Smithsonian Institution

On August 10th, 1846, the United States Congress passed the legislation 9 Stat. 102, creating the Smithsonian Institution as a trust instrumentality of the United States. It was to be an establishment dedicated to the "increase and diffusion of knowledge," and President James K. Polk signed it into law the same day. The legislation was the culmination of over a decade of debate within the Congress and among the general public over an unusual bequest. When the English chemist and mineralogist, James Smithson, died in 1829, he left a will stating that if his nephew and sole heir died without heirs, his estate should go to the United States to found in Washington, under the name of the Smithsonian Institution, an establishment for the increase and diffusion of knowledge among   men.

The Regents Room in the Smithsonian

The law established a Board of Regents which is responsible for the governance of the institution.   It consists of the Chief Justice, Vice President, three United States Senators, three United States Representatives, and nine citizens.

The following explains why the Smithsonian Institute tried so hard to ensure that their third Secretary, Samuel Langley, was credited as inventor of the first manned flying machine:

From 1903 forward to 1914 the Smithsonian argued that the Langley flying machine was the first one that would have carried a man if it had not been for an accident. The Institute paid Glenn Curtiss to restore the

1903 Langley Aerodrome, and he did fly it from Lake Keuka ostensibly to prove the Aerodrome was the first airplane capable of manned flight. In response to questioning by reporters, Curtiss admitted, however, that he had made 96 changes to the Langley machine before it actually would fly.

The Curtiss Rebuilt Langley Machine
This plane was flown by William Doherty over Lake Keuka in 1914.

In 1915 the court made a final ruling on the Wright-Curtiss patent. The Wright brothers were declared the rightful owners of the first flying machine to carry a man. The Smithsonian then went silent on the issue but did not admit that they were wrong until 1942 during World War II.

The Wright Brothers' First Flyer

Because of the treatment given the Wright brothers by the Smithsonian after their first flight, the brothers took pleasure in sending their original flyer, on which minor repairs had been made, to a London museum. They made an agreement with the British that if the United States ever apologized in writing and admitted that they, the Wright brothers, were the first to fly, then the English museum would send the flying machine back to the United States.

In 1948 a celebration was held in the Smithsonian Castle when the Wright flyer was to be installed in place of the *Spirit of Saint Louis*. The Wrights' first flyer hung there until the new aeronautical museum was built on the grounds of the Smithsonian property.

The Smithsonian Aeronautical Museum

First Wright Flyer Now Hanging in the National Air and Space Museum (NASM)

William Randolph Hearst (1863-1957)

Second International Air Show: In January of 1910, William Randolph Hearst decided that the United States must have an air show to compete with the international air show held in France. Hearst sponsored the Second

International Air Show to be held on windy Dominguez Mesa, located not far north of today's Los Angles International Airport.

Benjamin Foulois

When Lt. Benjamin Foulois was detached from the regular army while at Collage Park, Maryland, and appointed as the first member of the brand new Corps of Aviation, he received orders to Ft. Sam Houston, in San Antonio, Texas, along with the first Wright flyer accepted by the Signal Corps. He had been given no flying lessons like two other Army lieutenants, but he was the only one who had to learn to fly his plane without being completely checked out.

Lt. Benjamin Foulois (1879-1967)

On March 9, 1910, Lt. Foulois made his first flight, and by September he had flown the plane on 61 practice hops. During this period, the Wright brothers sent flying instructions through the mail to Foulois whenever he needed advice on some aspect of becoming a pilot. By early 1911, the airplane was in poor condition, having been wrecked and rebuilt by Foulois on several occasions, and it was retired from further service. In the above photo, he's standing in front of his first flyer in San Antonio, Texas.

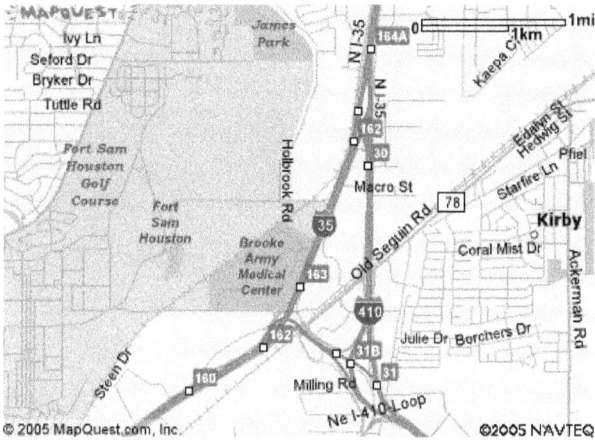

Fort Sam Houston, Northeast San Antonio, TX
(Now the Army Medical Center Division)

Lt. Benjamin Foulois
Fort Sam Houston, TX, in 1910

Benjamin Foulois joined the Army Corps of Engineers as a private on July 7, 1898, and spent most of his first years in Central America and the Philippines, rising to the rank of first sergeant in 1901. He was promoted to second lieutenant on February 2, 1901.

Lt. Benjamin Foulois and Orville Wright

In 1909

Memorial of First English Channel Crossing by Air in 1909

Placard in Front of Partially Buried Flying Machine at Dover (The craft has a weatherproof covering.)

Baroness de Laroche

The first woman to win her fixed-wing pilot's license was the self-styled Baroness de Laroche (real name Elise

Raymonde Deroche). Her photograph is shown earlier in this text.

Charles Stewart Rolls

Charles Stewart Rolls came from a wealthy family and was graduated from Eton with a Masters degree in 1902. Rolls, unlike most wealthy young Englishmen as pictured in American movies, had a reputation of being careful with his money, economical with food, and had a modest intake of alcohol.

Charles Rolls in his Wright Flyer.

Charles did like exciting things and had gone up in a balloon with his friend Moore-Brabazon before Moore-Brabazon became the 33rd pilot to fly a flying machine. Charles also liked high-speed autos and set a speed record in 1903 of 93 miles an hour, a world record at that time. In 1906 he established his famous automobile company Rolls-Royce Limited with a partner, Henry Royce.

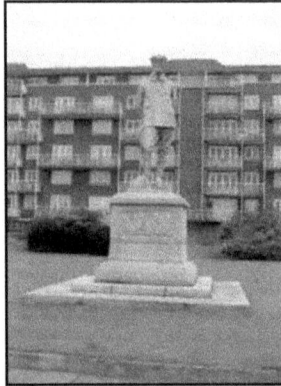

Statue of Charles Rolls at the seaport of Dover

On June 2, 1910, Charles Rolls, flying a Wright flyer he had purchased from the Wright agency in France, flew from Dover, England, across the English Channel over a town in France, and without landing, flew back to Dover and landed safely. He became the first pilot to cross the Channel in both directions. Several observers were posted at the town in France where he made his circle before returning to England. This was verified by the official observers.

On July 12, 1910, another international air meet was held at a town celebration at Bournemouth, England. Rolls was to participate by flying his newly modified Wright flyer at the air show. The Wright agency in France had developed what they thought was a newer and better tail for the Wright flyer. There is little information as to whether the French had actually tested the new tail that was fitted on the Rolls flyer.

After Rolls took off and climbed to several hundred feet, the newly attached tail broke completely off his plane, and the nose dived back to earth. Thus Rolls became the first pilot to be killed while flying an aircraft in

England. He was just 33 years old. This statute of Charles Rolls is located in the seaport of Dover and is near where he landed when he flew from England to France and returned.

John Theodore Cuthbert Moore-Brabazon

A good friend of Rolls, J. T. C. (John Theodore Cuthbert) Moore-Brabazon, the 33rd pilot to fly a flying machine, is credited with carrying the first live cargo in his Voisin, proving that "Pigs DO fly!" Note the pig is tied inside a barrel, which is tied to a strut. This photo was taken just before takeoff of the flight of the first live cargo in Great Britain in 1910.

This photo is of John Moore-Brabazon in his later years. He served in many capacities in the Government and was appointed by Churchill to the House of Lords

1909 Congressional Medals for the Wright Brothers

Three medals struck in France were presented to the Wright brothers for their many flying performances in France.

First Three Patent Applications

Wright Brothers

The Wright brothers made the very first patent application on a flying machine.

It was finally granted on May 22, 1906. If you want to know what a patent looks like, go to search engine Google and search for: *1906 Wright US Patent # 821393* (23 pages)

Dr. Alexander Graham Bell

Alexander Graham Bell applied for a patent called for as "A new and Useful Improvement in Flying Machines." This patent request listed 28 innovations that included ailerons, although they were not referred to by that name in the application. That patent was granted in December 1911.

Dr. William W. Christmas

William Christmas was the first pilot to fly a plane with ailerons at Fairfax, Virginia, on February 8, 1908. Dr. Christmas then applied for, and received, a patent on his ailerons.. His ailerons look much like the ailerons that are still used today. Below is a photograph of his airplane.

If you look closely along the trailing edge of each wing at the cutbacks on the rear of each wing, you can make out the actual ailerons. The photo below that flying machine is one of the planes built by the AEA, and their ailerons, which had a different name in Bell's patent application, are attached at the end of each wing.

First Flying Machine Built by Dr. William W. Christmas at Fairfax, Virginia

In the photo above you can see the actual wing cuts on the wing to your left where the ailerons were fitted along the trailing edge of the wing.

AEA June Bug Flown by Glenn Curtiss

In the photo above the ailerons are those triangles (I would call them winglets), and I am not sure what Bell called them, but he did get a patent on them.

Kitty Hawk Memorial

Kitty Hawk Memorial
80 feet tall

Orville Wright, Secretary of War Dwight F. Davis, and Amelia Earhart at the Dedication of the Memorial at Kitty Hawk

*1933: XB-15 Fly-by, competitor to the XB-17, on the 30th
Anniversary of the Wright brothers' first flight.*

Huffman Field

The name Huffman came from the owner of the field
who allowed the Wright brothers to use his vacant land to
develop their gliders and flying machines beginning in
1898. At the time of this writing, the field is set aside as
Huffman Park and is open to the public. It features
displays of Wright brothers' memorabilia.

It's located just adjoining Patrick Air Force Base in
Dayton, Ohio, southwest of the main Air Force Flight Test
Center. Huffman Park is where the Wright brothers first

tried out their gliders and flew their flying machines before and after 1903.

Those in the photo above are, left to right, Edward Warner, a founding member of the Civil Aeronautics Board (CAB) in 1938; Captain Kenneth Whiting, USN; Orville Wright; and General Hap Arnold. It was replicated and dedicated in 1938. This is the same memorial as it looks in the 21st century 2003. This memorial stands in a space between the Wright-Patterson Air Force Base and the entrance to Huffman Park.

The Wrap-up

This narrative encompasses the 6 years following the first flight of Wilbur and Orville Wright, America's most famous flyers. The highlights in their lives until the deaths of Orville Wright and Glenn Curtiss are briefly discussed in the following:

Glenn Curtiss, 1910 and Beyond

After the Herring-Curtiss Corporation went into bankruptcy and was dissolved, Glenn had three planes left. He paid his employees their wages out of his own pocket. and formed an exhibition flying team with the three airplanes. He trained two pilots, Lincoln Beachey and Charles Hamilton, to fly stunts and win pylon races and then transferred ownership of two of the planes to the two pilots so that they would have no liability under the patent laws.

Curtiss kept the racer that he had at Reims and managed to hang on to it, even though it could have been impounded under the patent laws at any time reputation was restored. Curtiss, Beachey, and Hamilton flew in close formation in much the same as the Thunderbirds do today.

The tangible assets of the Herring-Curtiss Company were sold at auction. The principal bidders were Herring and Curtiss by proxy. Twice Curtiss was the high bidder; twice Herring contested the result. After the third try, with Curtiss again the high bidder, he had bought back his company, and Herring was excised. He also bought the three planes mentioned in the previous paragraph.

A new organization, the Curtiss Motor Company, was formed and incorporated in December 1911 with Curtiss holding 50% of its stock.

The patent lawsuit continued until the final verdict came in 1913 that resolved the issue, and the Wrights won.

Curtiss took another shot at the lawsuit with unexpected help. That person was none other than Henry Ford, who had followed the litigation with great interest. Ford, with his battery of lawyers, had won a similarly difficult action with patents relating to the automobile, that was coincidentally heard by the same judge. Had Ford lost the case, his business would have been virtually destroyed.

This reopened patent lawsuit cost both the Wright Company and the newer Curtiss Company a great deal of money. Finally, the first verdict was upheld and the Wright Company was still the winner.

The Ghost Reappears

Just about the time that WWI began, Herring brought a lawsuit against Curtiss that accused him of secretly declaring bankruptcy of the Herring-Curtiss Corporation without his permission or knowledge. This suit dragged on for several years and was decided against Herring.

This should have been the end, but there was more to come! Somehow Herring managed to reopen the lawsuit in the mid-1920s, and it had not been resolved when he died at the age of 62 in 1927. Shortly after he died, the court reversed the first ruling and awarded Herring's widow $1 million to be paid by Glenn Curtiss. By that time, however, Curtiss was wealthy with an estate of about $30 million.

In 1916 Curtiss went public with his company as The Curtiss Aeroplane and Motor Company was moved to Buffalo, New York, and became quite successful. He soon built another plant in Toronto, Canada. Orders for several thousand planes came in and everything looked good. When the war was over, however, airplane sales dropped off and the company was losing money by 1920. A Canadian financer, Clement Keys, came to the rescue and obtained funds to manage the company.

The company then started building racing planes and won prizes in America and Europe for speed, altitude, and pylon racing. In 1923 and 1925 Curtiss' planes won the annual international prize for seaplane racers.

Jimmy Doolittle won the Snider Cup on October 26, 1925, flying the Curtiss R3C-2

In 1929 the Curtiss Aeroplane and Motor Company merged with the Wright Aeronautical Company to form the Curtiss Wright Company. Glenn Curtiss Died on July 23, 1930.

Modern Curtiss Wright Logo

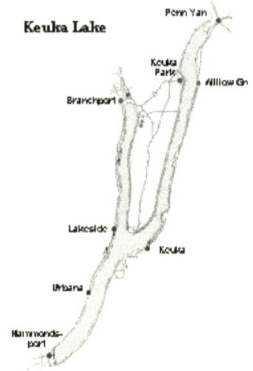

Hammondsport is the star in the red circle. New York City is in the lower right of the map. Curtiss flew flying boats, and float planes flew on this lake. Hammondsport is the town at the very bottom of the lake on the map.

Placard at the Edge of Field Where Curtiss First Flew His Aircraft

The Wright Brothers, 1909 and Beyond

If you are interested in knowing more about the Wright family after 1909, I recommend the book *To Conquer the Air* by James Tobin. The following paragraphs summarize the principal events that occurred after 1909 until Orville's death in 1938.

After Wilbur flew around the Statue of Liberty and finished training the two pilots for the U. S. Army at Collage Park, Maryland, near the end of 1909, Wilbur announced to his family and to the world that he would fly no more. He knew that although he and Orville had flown first, other pilots would fly higher, faster, and farther and would set new flying records in the future.

Wilbur changed his mind only once. On May 25, 1910, he rode as a passenger with Orville for the only time that they ever flew together. Wilbur climbed out of his seat and the two brothers talked their father, Bishop Wright, into flying as a passenger at least one time. So Bishop Wright went up for his first and only time as a passenger with his son Orville as his pilot.

While Wilbur ran their airplane business, Orville, like Glenn Curtiss, recognized that income could be derived from flight demonstrations, so he trained and flew with other pilots at fairs and airshows and brought in funds outside of airplane building. In 1915, however, Wilbur contracted scarlet fever and died on May 30, 1912. Orville then became the president and chairman of the board. Although he did not like administrative work, he kept the company going until he sold his interest in 1915. He was then a wealthy man.

For the rest of his life Orville was involved in aviation, flying, advising others about flying, and attending aviation functions. He lived with his sister, Katherine, who managed the household for the two of them. She had never gone back to teaching after she had left to care for Orville after his near-fatal crash. Their father came to live with them until he died in 1917, and then the two of them lived in the same old house. Orville started to build a new house on a lot he had purchased, but the house was never built.

Katherine had graduated from Oberlin College located in northern Ohio before she became a teacher. While she was in college she had been courted by a man named Henry Haskell but they did not plan to marry because she wanted to be a teacher and in those days, female teachers had to sign a contract that they were single and would not marry during the term of their contract.

So Henry started going with Kate's best friend and married her. Henry Haskell took a job that was offered on the *Kansas City Star* and eventually became the editor. Katherine and her girlfriend continued a lifetime correspondence, and when she died, Katherine attended the funeral. Then Henry and Katherine began their own correspondence.

In about 1925 Henry asked Kate to marry him. She thought about it seriously, but because she was afraid of Orville's temper, she refused to marry. Finally after about a year of wooing by letter on the part of Henry, Kate broke down and agreed to marry him. Kate was 50 years old when she made this decision.

Orville hit the roof, so to speak, and he refused to give her his blessing and would not attend her wedding. She moved bag and baggage to Kansas City to move in with her new husband, Henry. In 1929, on the eve of a trip that Henry had planned for them to go to Europe, Katherine contracted pneumonia and went to the hospital. Henry notified Orville that she was on her deathbed, but it took Orville's brother Lorin to convince Orville that he should go right away to see his sister at least once before she died.

Orville finally went to the Kansas City hospital and when he saw her, he did ask for forgiveness, which she gladly gave and then she died peacefully on that same day.

Orville spent the rest of his life in association with aviation events until his death in 1938.

Dedication Ceremony in 1948

On December 17, 1948, on the 45th anniversary of the first flight by the Wright brothers, there was a meeting in the Smithsonian Castle. Shortly after that gathering, the *Spirit of St. Louis,* Lindbergh's famous plane, was replaced with the original Wright flyer, which had been returned from England as promised.

Original Building and a Photo of Meeting Inside the Central Entrance in 1948

At this dedication ceremony of the original flyer that was to be hung in the place of honor, the brothers' oldest nephew, Milton Wright, was invited to describe what his uncles had done:

> *The Aeroplane means many things to many people. To some it may be a vehicle for romantic adventure or simply quick transportation. To others it may be a military weapon or means of relieving suffering. To me it represents the fabric, the glue, the spruce, the sheet metal, and the wire which put together under commonplace circumstances but with knowledge and skill, gave substance to dreams and fulfillment to hopes.*

ABOUT THE AUTHOR

First Squadron VMF - 911
Cherry Point, N.C. 1946

Twin Engine F7F-1

Official USMC   Photograph

Clyde Edwin (Ed) Middleton was born in Topeka, Kansas, in 1924. He attended public schools in Tulsa, Oklahoma, and was in his senior year of high school in Joplin, Missouri, when he left to join the Marines in November 1941 at age 17.

His career as a Marine spanned 20 years. He served as a radio operator, backseat gunner in dive bomber SBDs, and became an enlisted pilot, trained as a fighter pilot. He has had more than 8,000 hours of logged flight time in all types of aircraft, including jets. After his retirement from the Marines in 1961, he earned a Bachelor of Science degree in physics and mathematics from Memphis University, graduating in 1964. He then became a flight controller at NASA in Houston, Texas.

Upon the death of his wife, Fran, in 1972, he retired from NASA and moved to Laguna Beach, California, where he began still another career, this time in real estate. He married his second wife, Jean, in 1973. During his 27-year career as a real estate broker, he headed two real estate corporations.

He retired from real estate in 1997, sold his Laguna Beach home, and moved with his wife to Air Force Village West, a community for retired military officers in Riverside, California. He is now writing his memoirs and took time out to research and write this book on early flying machines because he felt this was a story that needed to be told.

It is his most sincere hope that you enjoy these things he thought you should know.

BIBLIOGRAPHY

Angelucci, Enzo. *The Rand McNally Encyclopedia of Military Aircraft 1914-1980*. New York: Military Press, 1983

The Best of Flying. Compiled by the editors of *Flying Magazine.* New York: Van Nostrand Reinhold Co., 1977

Boyne, Walter S. *The Smithsonian Book of Flight*. New York: Crown Publishers, 1977.

Bryan, C.D.B. *National Air and Space Museum* (second edition). New York: Henry A. Abrams, Inc., 1988.

Google Search Engine. http://www.google.com (images and general searches).

Jackson, Robert. *The Encyclopedia of Military Aircraft.* Paragon: 2002.

Taylor, John W. R. and Kenneth Munson. *History of Aviation.* New York: Van Nostrand Reinhold Co.,1977.

Toben, James. *To Conquer the Air*. Free Press, Division of Simon & Schuster: 2003.